The Best Book of

Speed Machines

Ian Graham

KINGFISHER

NEW YORK

Contents

Created for Kingfisher Publications Plc
by Picthall & Gunzi Limited

Author: Ian Graham
Editor: Lauren Robertson
Designer: Floyd Sayers
Editorial assistance: Barnaby Harward
Illustrator: Mark Bergin

KINGFISHER
Larousse Kingfisher Chambers Inc.
80 Maiden Lane
New York, New York 10038
www.kingfisherpub.com

First published in 2002
10 9 8 7 6 5 4 3 2 1

1TR/0102/WKT/MAR(MAR)/128/KMA

LIBRARY OF CONGRESS CATALOGING-IN-PUBLICATION DATA
has been applied for.

ISBN 0-7534-5436-X

Printed in Hong Kong

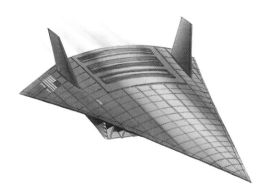

Fast, faster, fastest

Many people enjoy traveling at high speeds, but no matter how fast we go someone wants to go faster. In France in 1769 Nicolas Joseph Cugnot put a steam engine on a three-wheeled cart and made the first vehicle that could move by itself on land. The cart was so slow that someone walking could easily pass it. Since then people have built faster and faster cars, planes, boats, and trains. Some vehicles can even travel faster than sound!

Cat-Link

In 1998 a ship called *Cat-Link V* crossed the Atlantic Ocean faster than any other passenger ship. It took 68 hours to sail from the United States to England.

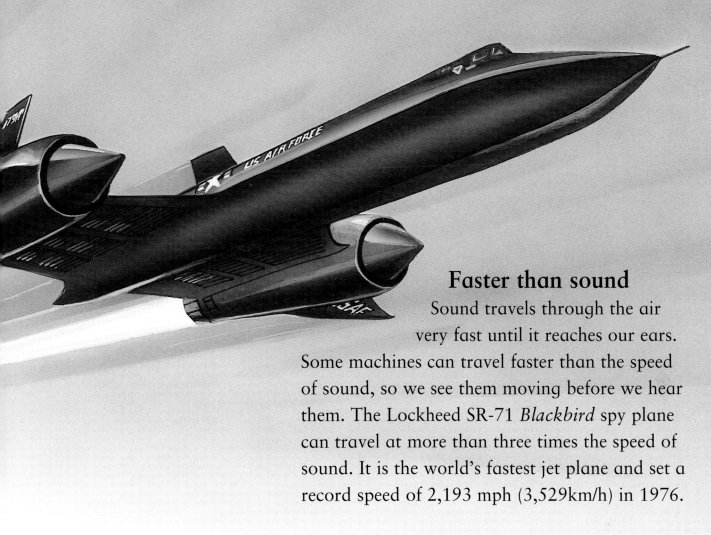

Faster than sound

Sound travels through the air very fast until it reaches our ears. Some machines can travel faster than the speed of sound, so we see them moving before we hear them. The Lockheed SR-71 *Blackbird* spy plane can travel at more than three times the speed of sound. It is the world's fastest jet plane and set a record speed of 2,193 mph (3,529km/h) in 1976.

The *Blue Flame*

In 1970 Gary Gabelich drove a rocket-powered car called the *Blue Flame* to a record-breaking speed of 630 mph (1,015km/h)—faster than a jumbo jet!

Sabre jet

In 1955 the world's fastest aircraft was an F-100C *Super Sabre* jet fighter. In that year it set an air speed record of 822 mph (1,323km/h).

Mallard

In 1938 the *Mallard* became the fastest steam locomotive by traveling at 125 mph (201km/h) while pulling seven carriages.

Stanley Steamer

In 1906 a steam-powered car called the *Stanley Steamer* set a land speed record of 127 mph (204km/h).

Bluebird

In 1964 Donald Campbell set a land speed record of 403 mph (649km/h) in his famous car, *Bluebird*.

X-15

In 1967 an X-15 rocket plane reached 4,520 mph (7,274km/h)—nearly seven times the speed of sound!

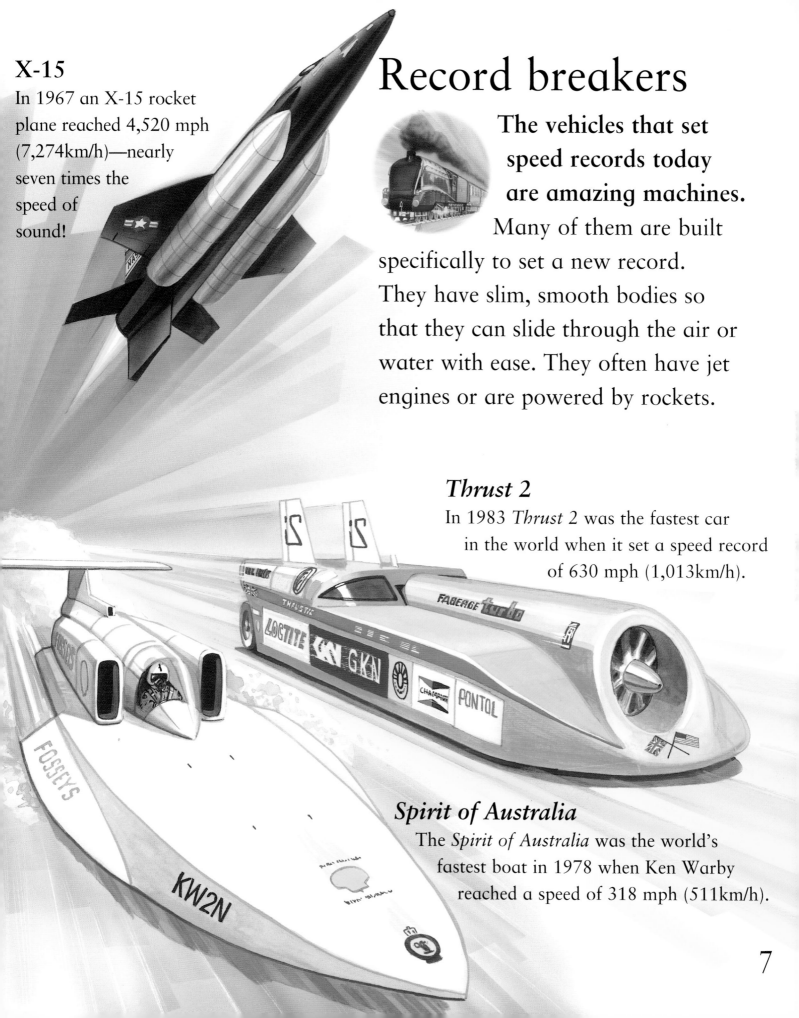

Record breakers

The vehicles that set speed records today are amazing machines. Many of them are built specifically to set a new record. They have slim, smooth bodies so that they can slide through the air or water with ease. They often have jet engines or are powered by rockets.

Thrust 2

In 1983 *Thrust 2* was the fastest car in the world when it set a speed record of 630 mph (1,013km/h).

Spirit of Australia

The *Spirit of Australia* was the world's fastest boat in 1978 when Ken Warby reached a speed of 318 mph (511km/h).

7

The land speed record

Of all the world's speed records, the one that people most want to beat is the land speed record.

In 1898 the world's fastest car had a top speed of 38.4 mph (62km/h)—slower than some roller-coaster rides today! By the 1960s the land speed record was 390.6 mph (630km/h), and a new type of engine was needed to go faster. In 1963 a new record was set using a car powered by a jet engine. The age of the jet car began.

Goldenrod

Bob Summers set a record of 407.96 mph (658km/h) in his *Goldenrod* car in 1965. It was the last record set before jet cars took over.

Rolls-Royce *Bluebird*

In 1935 Malcolm Campbell set his ninth land speed record of 300 mph (484km/h) in his car, a Rolls-Royce *Bluebird*.

Spirit of America

Craig Breedlove's latest *Spirit of America* jet car lost the race to set the first supersonic land speed record in 1997.

SPEEDVISION
NETWORK

JET

Shell Chemical

Duralast

Ford

GOODYEAR

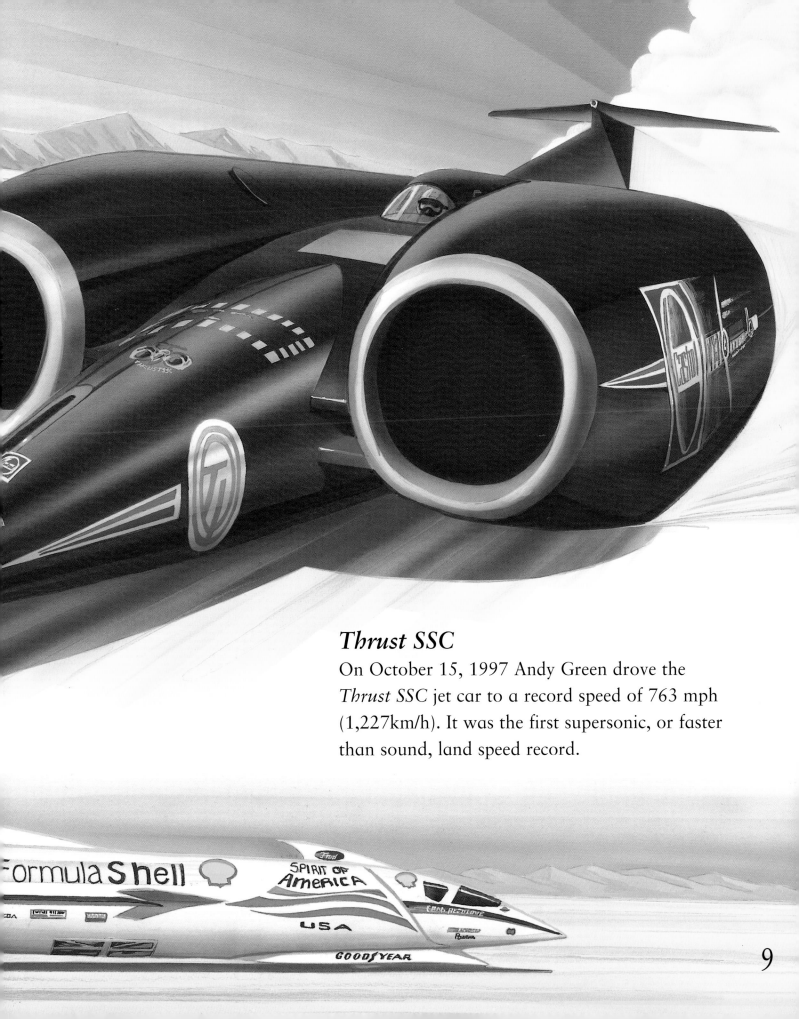

Thrust SSC

On October 15, 1997 Andy Green drove the *Thrust SSC* jet car to a record speed of 763 mph (1,227km/h). It was the first supersonic, or faster than sound, land speed record.

9

Winner takes all

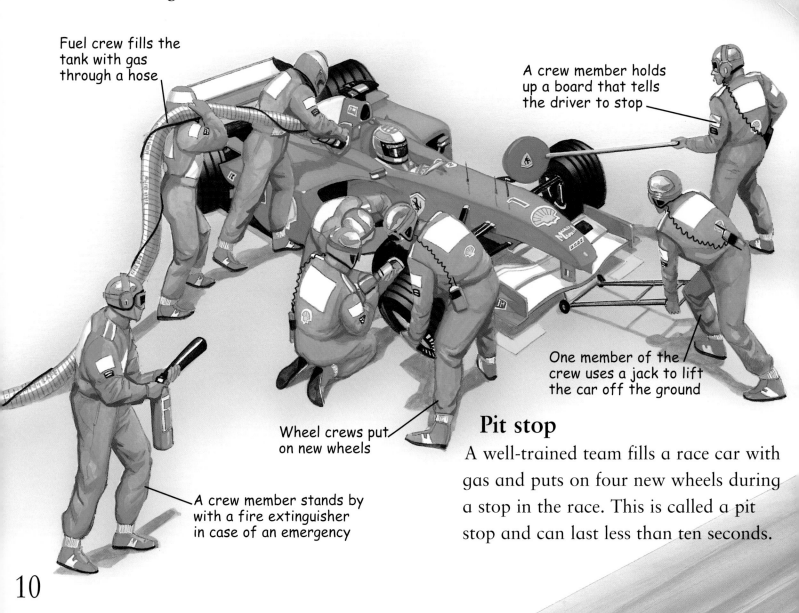

Motor racing is an exciting sport to watch. Most Formula 1, or Grand Prix, races are held on special tracks called circuits. The winner of each race is the first car over the finish line. Some races, called rallies, use roads through parks and forests. Ordinary cars cannot use these roads during the race. The fastest race cars are built low to the ground and have one seat in the middle of the car.

Fuel crew fills the tank with gas through a hose

A crew member holds up a board that tells the driver to stop

One member of the crew uses a jack to lift the car off the ground

Wheel crews put on new wheels

A crew member stands by with a fire extinguisher in case of an emergency

Pit stop

A well-trained team fills a race car with gas and puts on four new wheels during a stop in the race. This is called a pit stop and can last less than ten seconds.

Audi R8

Le Mans

The Le Mans sports car rally in France lasts for 24 hours, so each car has more than one driver. The winner is the car that travels the greatest distance in a fixed time.

Grand Prix

The marshal waves the checkered flag at the driver of the winning car. Waving this flag signals the end of the race.

11

What a drag

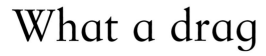

Drag racing is the fastest car-racing sport and is also the one with the shortest races. The fastest drag cars can reach speeds of more than 310 mph (500km/h), and a whole race can last less than five seconds from start to finish. Drag race cars, or dragsters, race two at a time on a straight track called a drag strip that is only 1,319 ft. (402m) long.

Dragster racing

In a drag race dragsters speed away from the starting line with a deafening roar of their engines. The dragsters' huge rear tires hurl them forward at top speed.

Tires that grip

Before a race dragsters spin their back wheels to heat the tires. This helps the tires grip the track.

Parachute braking

Dragsters use parachutes to slow them down at the end of a race. The parachute shoots out from the back of the car.

13

High performance

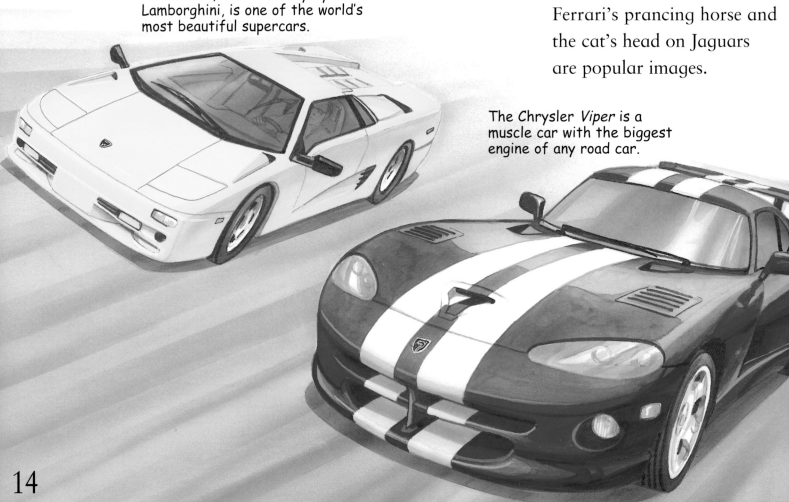

A high performance car is fast and fun to drive. Supercars and muscle cars are high performance cars and are the fastest vehicles on the road. Supercars usually have their engines behind the driver and are very comfortable inside. Muscle cars have very powerful engines, which are usually at the front. These road cars travel at speeds of up to 225 mph (360km/h).

Ferrari

Jaguar

Lamborghini

Lotus

Mercedes

Porsche

Insignia

All vehicles have an insignia that shows who made them. Ferrari's prancing horse and the cat's head on Jaguars are popular images.

The *Diablo*, made in Italy by Lamborghini, is one of the world's most beautiful supercars.

The Chrysler *Viper* is a muscle car with the biggest engine of any road car.

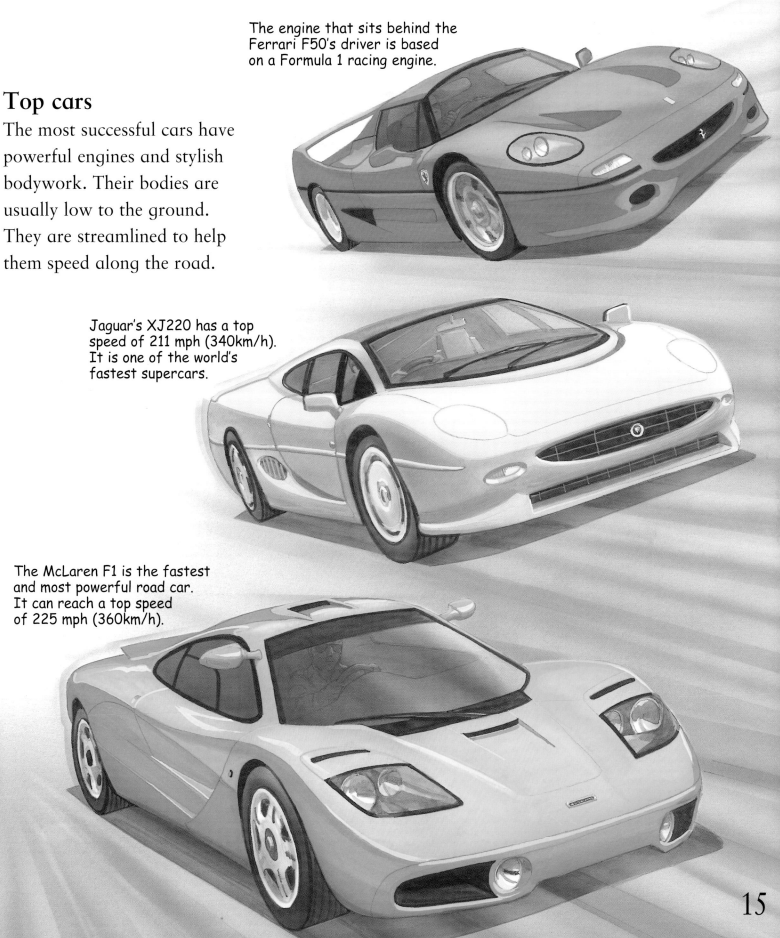

The engine that sits behind the Ferrari F50's driver is based on a Formula 1 racing engine.

Top cars

The most successful cars have powerful engines and stylish bodywork. Their bodies are usually low to the ground. They are streamlined to help them speed along the road.

Jaguar's XJ220 has a top speed of 211 mph (340km/h). It is one of the world's fastest supercars.

The McLaren F1 is the fastest and most powerful road car. It can reach a top speed of 225 mph (360km/h).

15

Pedal power

People have used bicycles for about 150 years, but we are still finding new ways of making them go faster. Many bikes have a frame made from steel tubing. Some of the fastest bikes use titanium, a lighter metal, instead. The most advanced sports bikes have a streamlined body made from carbon fiber. Riders wear tight-fitting suits and boat-shaped helmets so that they are also streamlined. In 1995 Fred Rompelberg made the world's fastest bicycle ride when he reached a speed of 166 mph (268km/h).

Mountain bikes

Mountain bikes are designed for riding on rough ground and steep slopes. They have an extra-strong frame and wide tires with extra-bumpy treads to grip loose and uneven surfaces.

Recumbent bikes

The rider of a recumbent bicycle or tricycle leans back with his or her feet at the front. This riding position is comfortable, but it takes a little getting used to. Fast bikes, like this one, are streamlined for speed.

Streamlined body helps this bike race along the road

Olympic cycling

Cycling is an Olympic sport. Two teams of cyclists start on opposite sides of an oval track and chase each other. This is called team pursuit cycling. The corners of the track are banked, or steeply sloped, so that the cyclists can race around the track at 37 mph (60km/h).

17

Two-wheeled speed

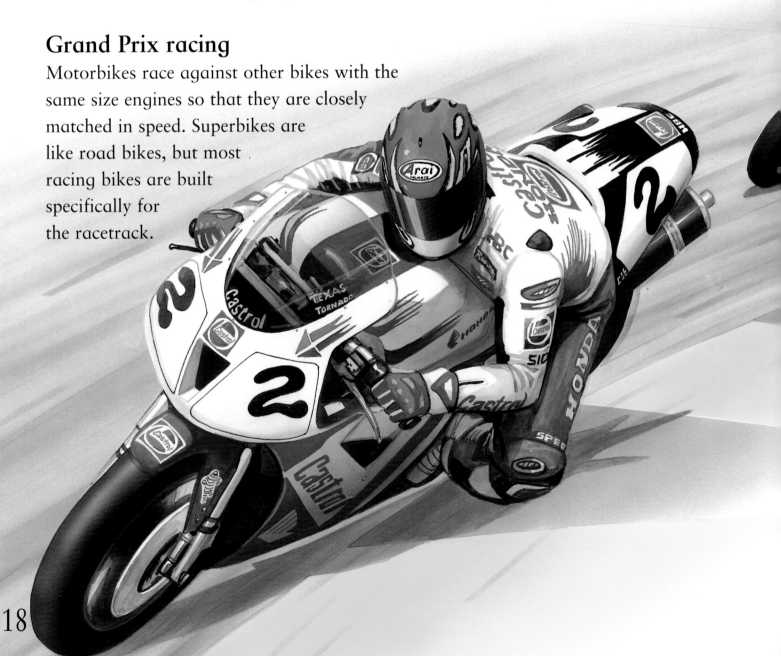

The fastest motorcycles are racing bikes that battle for victory on racetracks and are built to set speed records. The sportiest road bikes look like racers, and they are very fast. Anything that sticks out from a bike catches the air rushing past it and slows it down, so the rider has to crouch down and hug tight to the streamlined bodywork.

Grand Prix racing

Motorbikes race against other bikes with the same size engines so that they are closely matched in speed. Superbikes are like road bikes, but most racing bikes are built specifically for the racetrack.

Dragsters

Drag bikes are like drag cars, but with only two wheels. They have a monster-sized engine that powers a huge rear wheel. Drag bikes race two at a time down a 1,319-ft. (402-m) strip.

Supersports bike

The Suzuki GSX-1300 *Hayabusa* is one of the world's fastest road bikes. It can accelerate faster than a Formula 1 race car and can reach a top speed of more than 186 mph (300km/h).

On the rails

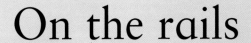

The fastest trains are powered by electricity from wires hanging above the tracks. In the future trains will be even faster and they may not have any wheels. There are trains being tested now that use magnets to make them hover above the track. They are called magnetic levitation trains, or maglevs.

Japanese Bullet

The first Shinkansen *Bullet* trains ran in Japan in 1964. They had a top speed of over 124 mph (200km/h). In 1997 an improved *Bullet* train with a top speed of 186 mph (300km/h) started making regular trips.

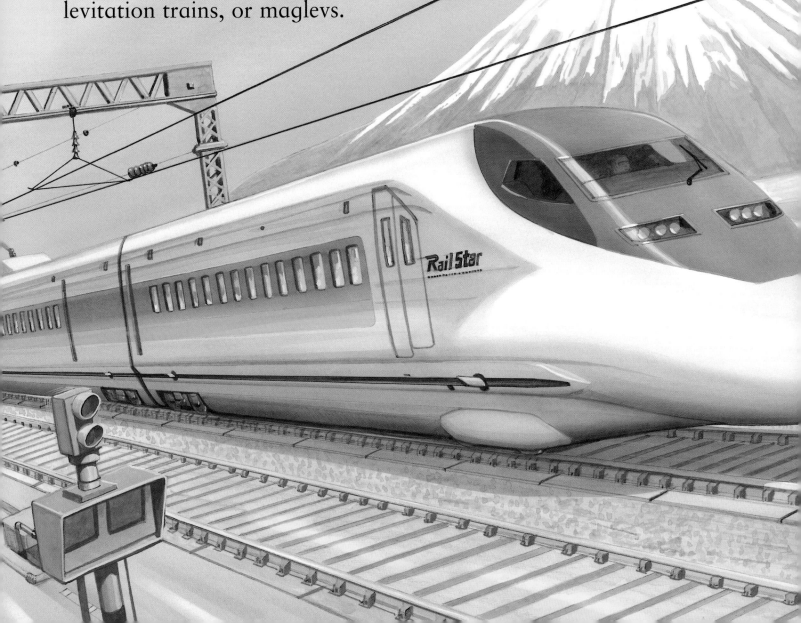

The ICE train

The German Inter-City Express, or ICE train, has reached a top speed of more than 250 mph (400km/h). It usually goes slower when it carries passengers, at 124–175 mph (200–280km/h).

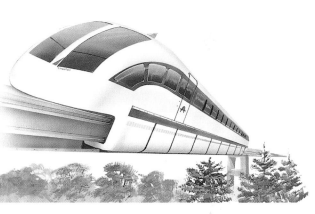

Transrapid

The *Transrapid* train is a German maglev. It hovers over a single rail called a monorail. It is not carrying passengers yet, but it has been able to reach a speed of 280 mph (450km/h).

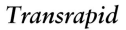

French TGV

The highest speed reached on any national railroad network is 320 mph (515km/h). This record was set by a French TGV *Atlantique* train between Courtalain and Tours in 1990.

World of water

For thousands of years boats had sails and could only travel as fast as the wind. Now boats also have engines, which move them much faster. But they are still slower than cars or planes because water is much thicker than air and more difficult to move through.

Jet skis

These machines speed through the surf by using a jet of water. An engine under the seat sucks in water and forces it out the back of the jet ski.

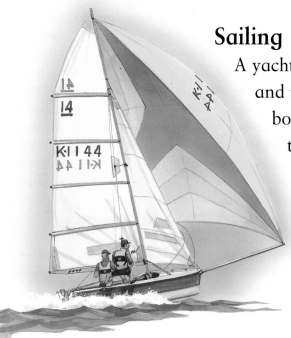

Sailing

A yacht's sails catch the wind, and this is what makes the boat move. The wind fills the sails and carries the yacht across the water. Sailors pull ropes to adjust the sails and get the most speed out of the wind.

Motorboats

Small, fast boats are the sports cars
of the water world. They can twist and
turn and are fast and fun to drive. Most
motorboats are driven close to the shore.
They are often used for trips around
lakes and for waterskiing.

Miss Freei

In 2000 Russ Wicks drove the hydroplane *Miss Freei* at a speed of 206 mph (331km/h). This was a record speed for a boat with a propeller.

Wave riders

Water is heavy to travel through so boats move slower than cars or planes. The fastest boats skim over the surface of the water. Only their spinning propellers dip beneath the water as they speed across the tops of the waves. Some of these wave-skimming motorboats have one long, narrow body, or hull. Others have two thinner hulls, side by side. Boats with two hulls are called catamarans, or multihulls.

Motorboat racing

Motorboats chase each other to the finish line of a race at up to 124 mph (200km/h). When they hit a wave at this speed, they leap into the air and slam down on the water again.

Fastest in the air

Airplanes that carry people are called airliners. Most of them fly up to 560 mph (900km/h). Some fighter planes can fly over 1,550 mph (2,500km/h). A few fighters reach speeds of more than 1,865 mph (3,000km/h). Fighters and most airliners have jet engines and are called jet planes. Before the first jet planes flew the fastest aircraft were fighters with propellers. Their top speed was about 370 mph (600km/h).

Air races

The fastest propeller planes from the 1940s are still flying today. Fighters such as the P-51 *Mustang* take part in the National Air Races held in the U.S.

Concorde

The fastest airliner in use today is the *Concorde*. It carries 100 passengers across the Atlantic Ocean at 1,367 mph (2,200km/h). That is more than double the speed of any other airliner.

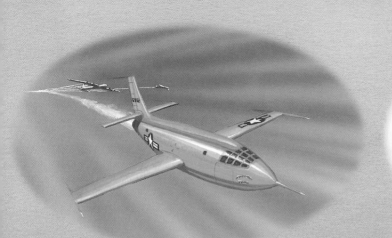

The Bell X-1

The first airplane to fly faster than the speed of sound was the Bell X-1. This rocket-powered airplane made its famous flight on October 14, 1947.

The MiG-25

One of the fastest airplanes in the world is the Russian MiG-25 fighter. Its speed has been measured at 2,110 mph (3,395km/h), which is over three times the speed of sound.

Going hypersonic

Only space planes, such as the X-15 and the space shuttle, have flown faster than about three times the speed of sound. Future aircraft will be able to fly at more than five times the speed of sound. This is called hypersonic speed. A small model called the X-43 is being used to test a new hypersonic plane.

Space shuttle

A space shuttle blasts off into space. When it begins its journey back to earth, it is flying at 16,777 mph (27,000km/h), or 25 times the speed of sound.

Testing X-43

The X-43 cannot take off on its own. It is attached to a rocket under the wing of a B-52 bomber. When the rocket is launched, it boosts the X-43 to 4,660 mph (7,500km/h). The X-43 comes away from the rocket, starts its engine, and flies on its own power.

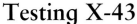

X-15

The X-15 was a plane powered by a rocket instead of a jet engine. In 1967 it reached a speed of nearly seven times the speed of sound.

29

How fast can we go?

On Earth's surface there are few places that are large enough or flat enough to drive a vehicle faster than we do now. Outside Earth's atmosphere there is no air to slow vehicles down. Speeds there can be much higher than on Earth. Aircraft built to go beyond the atmosphere can fly much faster than five times the speed of sound!

HyperSoar

This special craft will skip across the top of the atmosphere at ten times the speed of sound. It will be able to fly to anywhere on Earth in only two hours.

Apollo 10

The fastest speed that any human has traveled is 24,792 mph (39,897km/h). This is the speed reached by astronauts in the *Apollo 10* spacecraft as it returned to Earth from the moon in 1969.

Speed sense

Not every new aircraft is faster than the previous one. The French *Rafale* fighter can tumble and turn more quickly in the air than many older, faster planes.

Glossary

accelerate To go faster. The pedal in a car that is pressed to make the car speed up is called the accelerator.

aircraft A machine built to fly through the air.

airliner A large plane that carries passengers.

atmosphere The air that surrounds the Earth.

bodywork The external part of a car or bike that surrounds the engine and the passengers in a car.

bullet train The high-speed Shinkansen train in Japan. It is called the bullet train because of its speed.

carbon fiber A material that is lighter and stronger than metal. It is made from strands of carbon inside hard plastic sheets.

dragster A car, bike, or boat used for drag racing. These vehicles are designed to reach top speeds as quickly as possible along a straight line. Dragsters race on a track called a drag strip.

fighter plane An airplane built for the military to fight other aircraft in times of war.

fuel A material that is burned to make an engine work. Gas is a fuel.

hull The main body of a boat that sits in the water.

hydroplane A motorboat that has floats on each side of its hull. When the motorboat travels at very high speeds, the hull lifts out of the water and the boat rests on the floats at the front and the propeller at the back.

hypersonic A speed faster than five times the speed of sound. Five times faster than the speed of sound is about 3,355 mph (5,400km/h).

jet engine The type of engine used in setting speed records. It heats air so that it rushes out of the engine as a high-speed jet.

locomotive A railroad engine that pulls carriages.

maglev A magnetic levitation train that uses magnets to hover above a magnetic track.

parachute A large piece of fabric that opens out to catch as much air as possible. It is used to slow down someone parachuting out of a plane, to slow down a plane after it has landed, or to slow down a fast car.

rocket An engine that burns fuel to produce a jet of gas. Air is needed to burn fuel, but a rocket can work in space where there is no air.

speed of sound The time it takes for sound to reach us. The speed of sound is about 760 mph (1,225km/h) near the ground. High in the sky the air is cold and sound travels slower. Here the speed of sound is about 660 mph (1,060km/h).

streamlined Slim and smoothly shaped to move easily.

supersonic A speed that is faster than the speed of sound.

Index